School Publishing

ROCK TOUR

PIONEER EDITION

By Beth Geiger

CONTENTS

ROCK TO

Would you like to meet some real rock stars? The United States has amazing scenery, all made of rock! Let's explore some of these wild wonders.

BY BETH GEIGER

JR

Valley of Stars

Monument Valley, Utah, is not just for rock stars. It's full of movie stars, too. Dozens of movies have been filmed here.

Imagine a huge layer cake made of rock! Monument Valley is mainly a sedimentary rock called **sandstone**. Over time, water and wind nibbled most of the rock away. All that's left are fragile pillars and mountains with flat tops called **buttes**.

Monument Valley has a special magic to it. There's no place else quite like it.

Underground Wonders

A glittering palace hides under the New Mexico desert. It is a cave called Carlsbad Caverns.

Step into Carlsbad Caverns's Big Room. It's as big as 6.2 football fields! Rock formations hang from the ceiling. They look like icicles.

Carlsbad Caverns formed from a rock called limestone. Limestone is a type of sedimentary rock. Millions of years ago, water flowed here. The water dripped and flowed. It combined with acidic water and gases. Then it dissolved the limestone. Slowly, Carlsbad Caverns's formations took shape.

Meeting in the Middle. Stalactites and stalagmites form columns where they meet.

Tree Tales

Once, Arizona had a lush forest. Dinosaurs roamed there. The dinosaurs disappeared. The trees fell.

You can still see logs from that ancient forest. But they aren't wood anymore. They are rock. When wood turns to rock, it's called petrified. So this place is called the Petrified Forest.

How does wood turn to rock? After the trees fell, they were covered with ash from a volcano. Slowly, minerals from the ash replaced the wood.

Ship in the Desert

A ragged mountain looms above the New Mexico desert. This pile of **igneous rock** is called Ship Rock. Ship Rock didn't always stick up. Once, it was an underground pipe filled with melted rock called **magma**. It was part of a fiery volcano.

The volcano erupted. The leftover magma in the pipe cooled. It hardened. Slowly, everything around it weathered away. The thick pipe was left sticking up into the sky, like a ship on a sea.

Tall Tower. Ship Rock is 600 meters (1,700 feet) high.

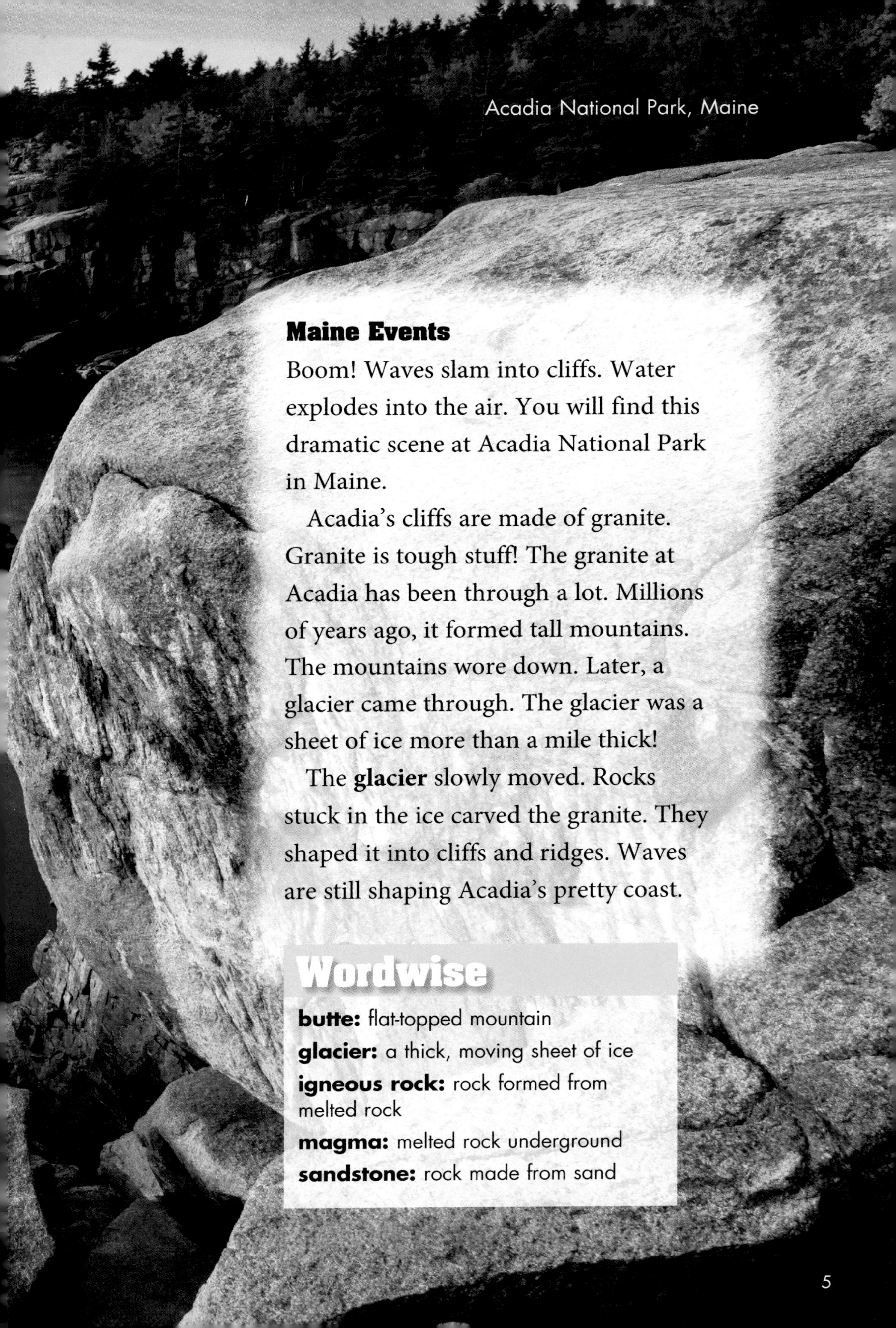

Acadia National Park, Maine

Maine Events

Boom! Waves slam into cliffs. Water explodes into the air. You will find this dramatic scene at Acadia National Park in Maine.

Acadia's cliffs are made of granite. Granite is tough stuff! The granite at Acadia has been through a lot. Millions of years ago, it formed tall mountains. The mountains wore down. Later, a glacier came through. The glacier was a sheet of ice more than a mile thick!

The **glacier** slowly moved. Rocks stuck in the ice carved the granite. They shaped it into cliffs and ridges. Waves are still shaping Acadia's pretty coast.

Wordwise

butte: flat-topped mountain
glacier: a thick, moving sheet of ice
igneous rock: rock formed from melted rock
magma: melted rock underground
sandstone: rock made from sand

ROCKIN' RECIPES

Rocks are like cookies. They are made of different minerals and parts. Oatmeal raisin is just one type of cookie. Each rock shown is also just one type of sedimentary, igneous, or metamorphic rock.

Oatmeal Raisin Cookies

These cookies are made from different things, just like rocks.

Sedimentary Rock

Conglomerate is a type of sedimentary rock. It's made of small pieces of other rocks.

Granite is a type of igneous rock. It forms from minerals interlocked together.

White/gray quartz

Pink feldspar

Metamorphic Rock

Schist is a type of metamorphic rock. Its bends formed from pressure and heat.

Red garnet

Green chlorite

STAMPEDE!

It was a winter morning in 1848. A man worked near a creek in California. Suddenly, he spotted something bright. Gold! Sure enough, the creek was full of gold.

The news spread fast. People raced to California from all over the world. The great Gold Rush was on!

The 49ers

Many of these people came in 1849. So they were nicknamed 49ers. The lucky ones found gold. Some bits of gold were as small as dust. Others were as big as your fist—or even bigger.

Later, many 49ers stayed in California. Towns and cities grew. The Gold Rush changed California forever.

Golden Cities. Cities like San Francisco grew from the gold rush.

Treasure of the Sierra Nevada

California's gold comes from a rock called granite. The granite forms a tall mountain range, the jagged Sierra Nevada mountains.

The Sierra Nevada mountains have been slowly crumbling for a long, long time. Rain and frost wear them away. Bits of broken granite tumble and wash into streams. The gold washes into the streams along with the granite.

Sierra Nevada Mountains

Mining for Gold. People probably found gold like this in Coloma, California, where the first gold was discovered.

Sinking Nuggets

Gold is heavy. In streams, it sinks right to the bottom. Gold doesn't crumble. Bits of it stay whole, like raisins in a cookie.

Gold collected in California's creeks for millions of years. Bright yellow flakes and nuggets of gold sank into pools. They stuck in cracks. Then, the 49ers came along.

Hard Work

To mine the gold, 49ers scooped up heaps of gravel. Then, they ran water over it. The gravel washed away. The heavy gold stayed put.

Life in the gold camps was hard. Sure, most 49ers found some gold. But few got rich. They spent their money on costly food and supplies.

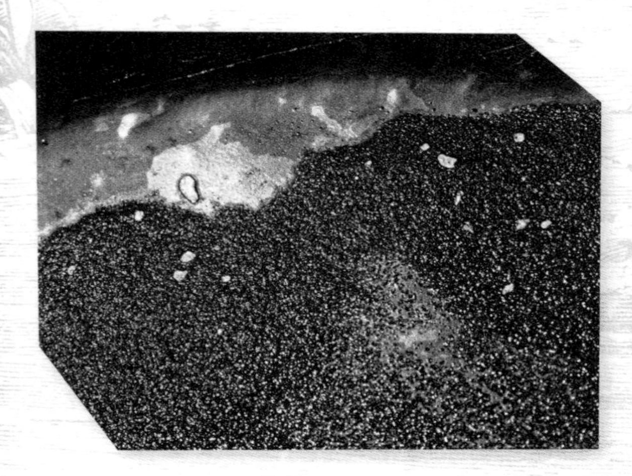

Cleaning Up. Heavy gold is left behind when you wash away rocks and soil.

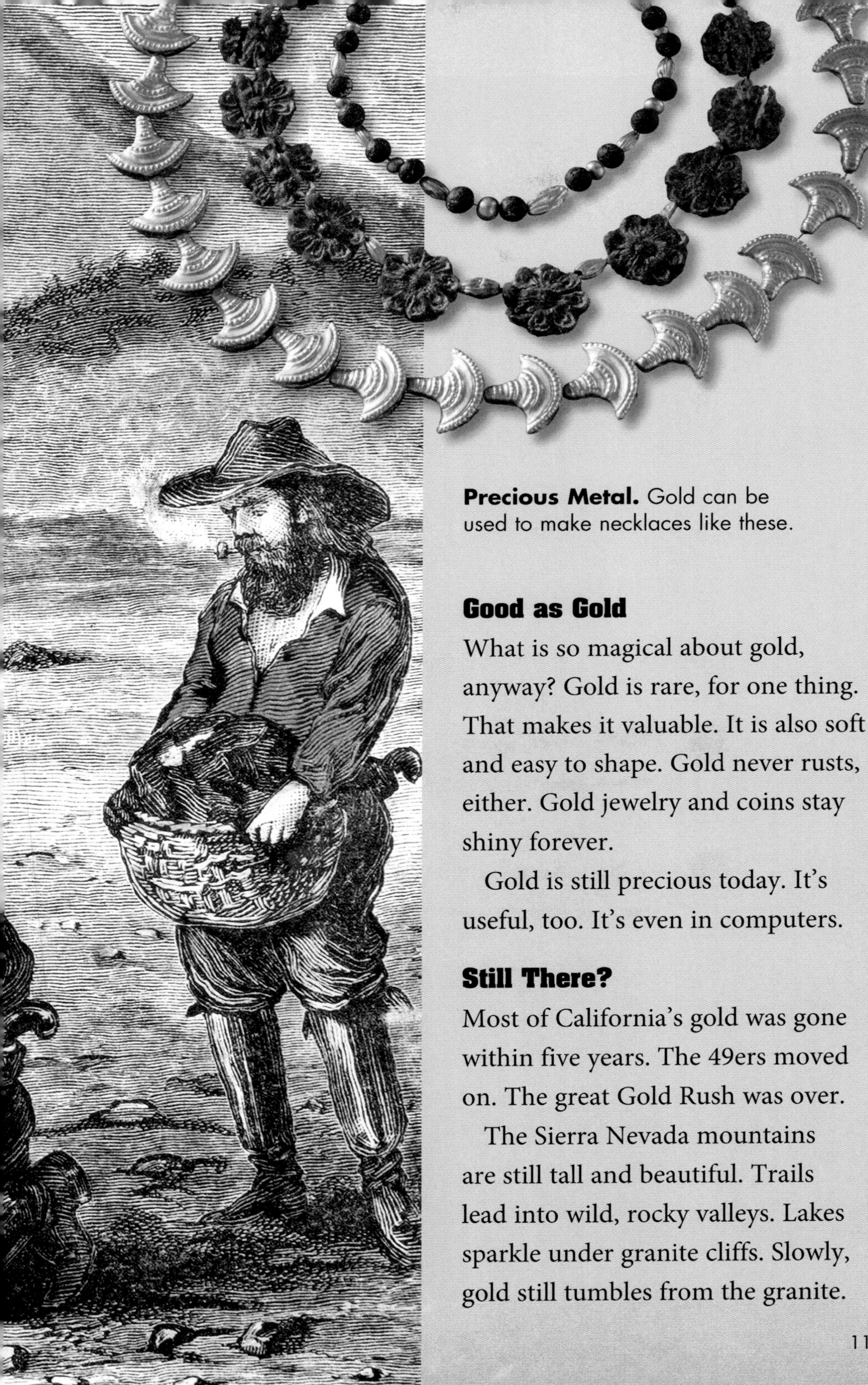

Precious Metal. Gold can be used to make necklaces like these.

Good as Gold

What is so magical about gold, anyway? Gold is rare, for one thing. That makes it valuable. It is also soft and easy to shape. Gold never rusts, either. Gold jewelry and coins stay shiny forever.

Gold is still precious today. It's useful, too. It's even in computers.

Still There?

Most of California's gold was gone within five years. The 49ers moved on. The great Gold Rush was over.

The Sierra Nevada mountains are still tall and beautiful. Trails lead into wild, rocky valleys. Lakes sparkle under granite cliffs. Slowly, gold still tumbles from the granite.

ROCK TOUR

Answer these rockin' questions to see what you've learned about rocks.

1. How did the wood in the petrified forest turn to rock?
2. Why does Ship Rock stick up into the sky?
3. How does igneous rock form?
4. How did gold get into the streams near the mountains?
5. Why does gold sink down to the bottom of streams?